Poems and Sketches of Bipolar Schizophrenia

By Tristan B. Gilbert

Poems and Sketches of Bipolar Schizophrenia

By Tristan B. Gilbert

Published by Tristan B. Gilbert

International Standard Book Number: 978-0-6151-7275-0

TristanGilbert.com

Doorway to the Independent Music Revolution

Bend your mind into the realm of mental illness with these thought provoking books written by a seriously disturbed individual. Tristan Gilbert for many years rode the fine line between genius and insanity until he finally caved into his delusionous imagination and was institutionalized for his own protection. Regaining a new chemical balance in his brain through modern antipsychotics, Tristan shares his past, present, and in future books; his future with these artful masterpieces sure to marvel the intellectual world for many generations.

Every time a sit in lace
I praise the day
Of all the chase
Tender touch
Upon your lips
In the night
It is called a kiss
Touch me tender
Touch me true
In this life
I will always love you
Pamela in my Heart

Wiccan ways
Choose to stays
Mastermind
Sometimes blind
Others behind the wheel
Cannot say
Another day
Time after
Choosing
Not to direct
myself
In a lawful direction
Flitter Flatter
Extreme Tarrif
On pleasant dreams
Wanting to grow
Not to demean
But knowing
Shed a Light
Daunting
Get it Right

still standing
Banding}
My insanity
Of which I choose
Not to go without
A high cost
Of guilt

However
Must I promise dispeaR
Glare
For Law abiding citizenship

For it is cannibus
That I so much desiRe
Lady of my esteem

Rotten Moments
Rewritten
To kindler mindler
Direction

For the choosing
Floosing
Grateful eye
Heavens laughter
Not so shy
But confused
Chicken slaughter
EveR AfteR
He so got her

Trembling along
So strong
Belong

Volume High
Along the wave
Life surfing
In a maze
Haze lifting

Lusting for belongings
Of what I do see
Cheerful imitations
Getting through a day
Stop
Beleiving

Steeped with promises
Never ending in truth
Belched between words
Fragmented
Fever laden
Daunted
Not Dead
BLANDED
Some Red
Flitter
Flatter
Over Run
In Pleasure
Lusting
Without End

WORD
of
Mouth

In Search of
Holy Grail
Flowers
Blooming
this day
NOT
Satan
Hades
On you tail
Torn
On
The Edge
Facing
Inspiration

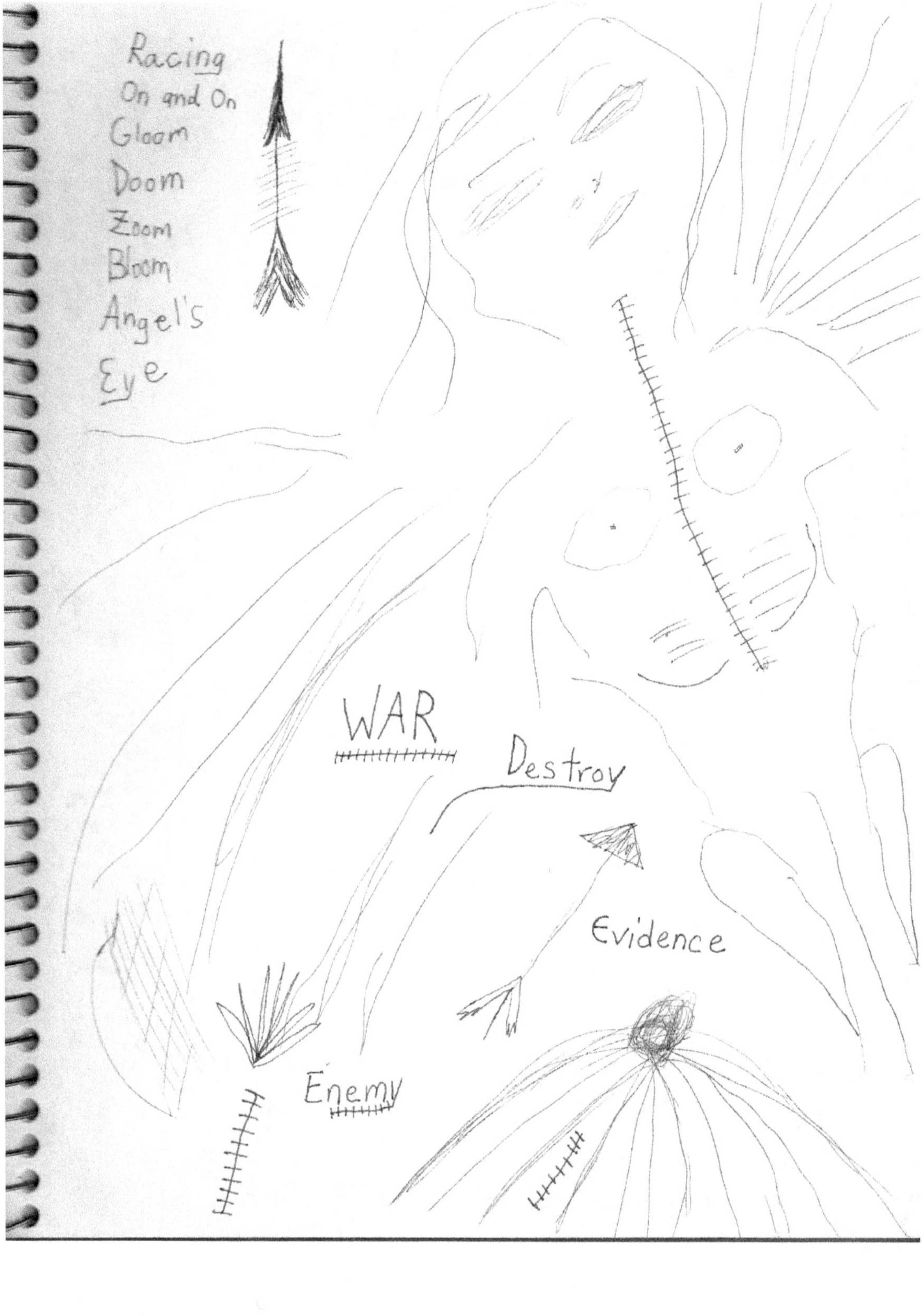
Racing
On and On
Gloom
Doom
Zoom
Bloom
Angel's
Eye
WAR
Destroy
Evidence
Enemy

Blinded
In a nutshell
Away from the doom
Of Satan's laughteR
There is no room
Let me stay
Time to play
Knocking on heaven's dooR
Rainbow
treasure
wanting
with
Lustful nature
Maiden
of
Vision

Time will tell
When all hell
reeks upon the floor
Let me in
Up to chin
With feces upon the mind
Dirty Laundry
Wanting to see
Error
which permeates the mind
Hallucination
Blasting on
Timeless platter
From which to feast
The Beast
of mind control
When I Am you are young
Think adventure is trying new things

Oldness is regularity
Amskray as lightening
With a Rod
Thinking God
Into the darkness
Facing hardness
Dowsing into the maddness

Time sits still and I chill

Thoughts meandering in every which direction

Laughter pon Laughter some giggling

Childesh seasoning

chicken soup

Wantingly Wondering

Fuck lips in her pants

Lustingly queeving

How best to have her

On and On I wonder

Til point of bearing

Or longer without

chicken fisted torment

Hah Hah

Snod Grass

Burn without excitement

Hurling matter

In odd direction

Smearing snot

Bleakenly

Trump Card

Moment dismilly spoken
Ace in a hole
Groucho Marx on the spot
Hello
I Am here
Lost and bewildered
Without proper language
To explain my position
We are like animals
Death is upon us
In circles
Buried in insignificant achievement
Trophies
Best in show
Mentally disbarred of organised thought
Perfering mental states
Not of government approvement
Stoned
Fryed
Memories
Nightmarish And KindeR

Wondering in Marryment
How can it be
A Long yondeR Fantasy
Nearing achievement
Flittering demised
Anarchy
Takes a great leaders
To defeat Non anarchist forces
Bill of Rights
Anarchist Thought
Right to bare arms
Short sleeve shirt
Wandering
Unfocused
Searching for Freedom
Abolishingly
Night rideR
Open sky
Fly By
Shooting

Torn from Reality

Quivering
Thoughtlessly
Stretching
To gain motion
Extended
Yoga like
Anarchy under Bill of Rights and ten commandments
Gridlock
So slow
Down the hill
Around the Bend
Just Right Here
Try to blend
Surrounding sun
Block
Kick and recieve
Believe
In irrigation
Whats to percieve

No blocks
Boarders more open
Free trade
Blade for Blade
Harmonious numbers

None to Raid

GRID LOCKed
opening

Traffic Jam

Hysterical Concourse

Blam Man

HELL

No one knows what is to be

Vaughter
Mercy dominion

Horror

Damnation

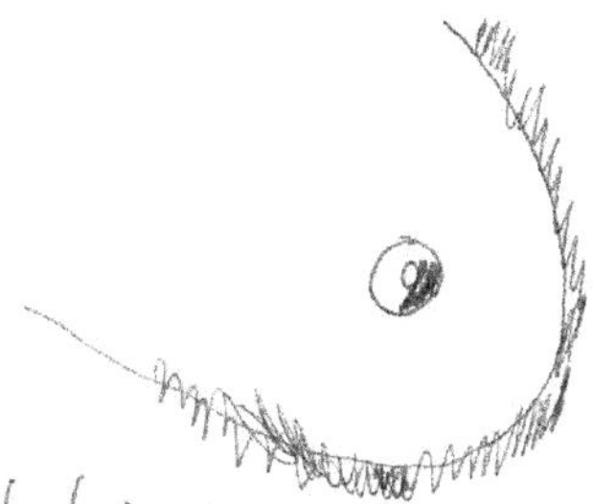

Devilish
Delight
Female
insight

Fantasies
Afflicted
With
demonic
Passion

Gates
Close in
No one to
win

Thinking of problems to solve
Not realy caring
Alone
In my mind
Visitors gone
On my own
Despair
Thoughts
Infiltrated
Stuck
Wanting out
Blaming none
TAMED
Dying
Thinking
Of a Dream
Seeming
Lost
AND
Looking
For a place to be
Crazy and Free

Houses should Have alternate power sources
It is insane to be at the mercy of the power companies
Food

Earth Destruction

Poisoned
Damnation
Roaming Heart
Must be set free
Water but Needed wine
City of Rising Son
Rising AHEAD
Falling Down
Many pains of Sorrow
Manifestations gone astray
Bah Bah Sheep
looking for a Way
To and Fro
Want to Know
Standing Fast
Slowing down
Curves and promises
who to Know

Sunshine
Earlymorning dream
Snow upon a shelter
Something leaving
Damned aferthough
Violent grieving
Bearing Heavy burdens
Breathing slim alone freeness
Hungry Spirit calling love to me
Surise through the wind
I did cry let dim
Take me feaverishly upon
A sensation
Retention
Calm cool hit
Triumphantly Burning
After a Mass
Removed
With gust
dis me man ?

Take a turnigate

Through a Hill

Round about thrill

Stapping

Beyond shelter

Fountain of bliss
give me a kiss
Don't miss
The toad prince
prowles
Upon you heart

Sniff
fragrance Recieved
Primal instinct

Saturated in Nothing
Countessly clueless
Courteous
Humbly Sniped

Sandwiched between promises
Not withstanding
Feverish moments
of thought
Estarcys
Stop blap
In a soda due
Forgotten again
Helping out
Wondering
If this thought moment percists
I will surely die
But I will surely dye
All most die
And Be recreated
Flattened in esteem
Without means
Sortless misowning

Mainly arbitration
Flitter Flatter
Grand illosieR
All and all of which
Seems into Fantasy
Disturbing discontent
Stop
Disillusionment
Pot can never be legalized
Seperates good guys from bad guys
Ticking
Flaunting
subjectivity
Illaceration
Despensizing
Triangle cultivated
Dreaming
End to WaR

End to war illegalization
Soft maufia
Smoking in public citation
misery
In company
POT CLUBS
ENGROSSED
Anti alcoholism
Killing
Destroying
Ways wonder
Salty misory
Counter
Bottle
Bliss
Wanting

Transfering Now

The declaration of an international war over a plant cannibus is wholy unconstatution in nature. Farmicutical. BLACK Market

True competion

Opium extracts

Intrepid is a car

Breathing Efficiently

FIRE STORM

Nazi Rule

Facism

True Torrent

Must there be a counter War to Reach Balance

Against communist dogma

No Dogma shall stand

Anarchy will not exist until the name is forgotten

Sitting on A Park Bench
EYING LITTLE GIRLS
SNOTS WANDERING
I Oh so cute of curls
But they is rotten to the core
Soon to be a whore
Someones crabby little bitch
Someones head is sure to itch
Catch my little cooties
Burn em in a fIRE
They is sure a witch
Pagan Pot smoking cunts
The cooties FOR sure
Everything must die
Grasshoppers do not cry
Cricket cousin intestinal fluke

Covered in soot

Silence of the Lambs

Vision
Light patterns

Evil At Kind
OF Devils Demise
mysterius of untowhich hazard
Gods Blazen Blew
Bellowind Brandish thoughts
Dispairmently Knowing
Heavens of Kind
May Never Know
OF Earthly Feelings
At heavenly moments
Ancient Kind Created
Fost Fenderlossly Feasting
Barbeque sandwiches

Mammories
Leased

Pickled and twisted
As ungrateful moment
Of weaning

Tooling the hazards
of Feindessly feasting

Given a give
Will have to do
NEVER AGAIN

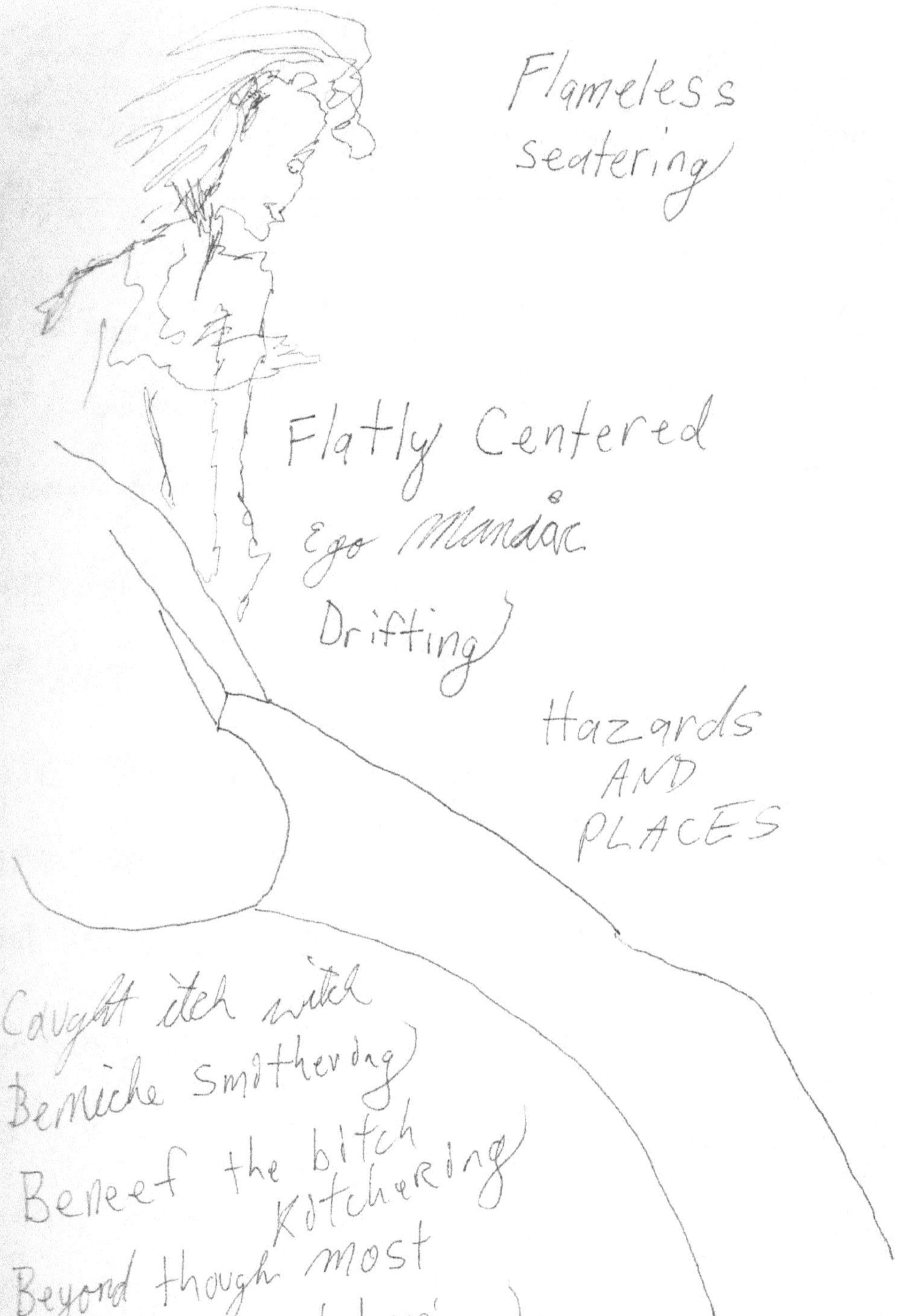

Flameless
seatering

Flatly Centered
Ego Maniac
Drifting

Hazards
AND
PLACES

Caught itch with
Bemiche smothering
Beneef the bitch
kotchering
Beyond though most
sutchering

Tainted and twisted
I Lost Fantasies Versian
Pittle pea a persian
Nightmarish LosteR
All Forgotten in the Sky
Jereasa promises of forgiven
WondeR Lost not be forgott
Lot not blisted
Kite in the sky
Damb do or dye
Kant tel a ly
Forgottenly friendly
End till I see
Faces Nawed and withered
Doom frim toweRED
Lost upon Freedom

Spirit alongside
Seated in Watching
Only interacting
To come + go

sword in my head
yet I live in
astonishment

People around keeping
such composure
Like they see one like
me everyday

But they are afraid

Never Has one like me ever
Been
Nor will there ever be

sword in my head
Yet I live

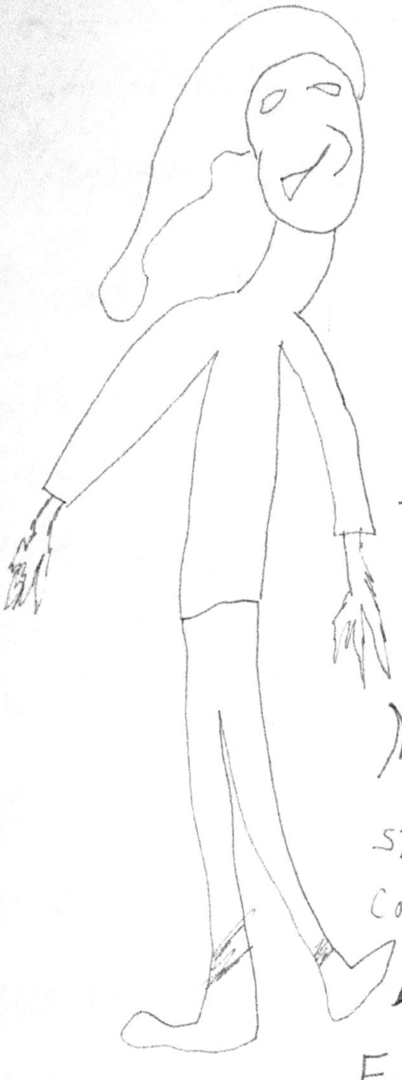

Faintly feeling forgotten
AS Flickering gently
Memories erased
Of A great man
By corrupting A Nun
Into short near damnation
To off through Balance
OF Goodness and evil
To let Satin out of
the Lair

Named borrowed stolin
stalin stiched Moments
Confronted in Nature
LATEly stolen
For heavens Laughter
To be taken

What now must I see ordue

To avoiding hazards of spirit

Razors Edge

Lats of Pledge

MASTERY of Say and due

Payment upon receipt

Death seems of evil kind

Never to remember

Moments of Peace

UPON the terror

Flashbacks continue

NOT WITHstanding

Further adue compensation

MeanderING

ManderlNG

Back off

quiet

story continues

Hells open as my Father is now Erased
No Nuns unto wich greet him
Wifes little better than a whore
Get what work ye can git
For little is known
OF Prevencés Geetished
Into demoshment
Obliterment of scottish
Foundlshly Feasting
Upon the flesh
OV A Bitch whore
Damb Blam Kill a cun
Blend her steep her
None to stir
Damb Blasted Morons

Some days I wonder
Wanting

Dreams of passion permeate my brain
Eternal laughter
Where gone the flame
Over ambition
No one would listen
So I hunted on
Something was missing
aberition of the past
Home at last
Can't be seen
lost in dream
Lay lie wandering
Where it must be
Tantilized + twisted
so strong gotten
Bested
In blemishes

Tiger wooded
Lost in the Jungle
Can't be none
So Long John
Mortition Awaits
Not Heavens gates
OR Even Bill Gates
But Hell on my heels
Devils own donkey
Can't wait
Fate of death
Could be better than Here
Die into better tomorrow
Politition
Flitting upon the flatter
Dance due us die

Turning upon a twisted do
Never of which lasted
Always ending
Change
coin time

phrasen idiots
Blasting past
Unburdened
wrapped
In
Toothpicks

Dismal desperate attempts
To Kill
Cause death
Through mutilation and suffering
Torture
All at once
For a long time
Sodomy
Suffication
Abuse
mame
Hit
Cut
Inject
Ravage
Hurt
Damage
Beyond repair
Total out

I think

Love is a never Ending gloom
For All to see
Lost in an adventure
Curling to a forever gene
Unaware of the surroundings
Selling out
Prostituting my very soul
To whoever wishes to recieve
What ever I have to give
Often Rotten
But choosing to surrender
For Life fantasy
of Forgiving
Natural ways of growth
To the End

Now

Dr. No

I want to Kill little white bunnies with bushy grey tails.

Chop off their heads and watch them bleed

From the biggining brings end
To End brings beginning
Clop top not rot but snot
In the early morning breeze
Hate the persuit
Live to hate all etistance
Bring Satan upon the land
So grand by hand
Crops merging endlessly
Where art thou heavely lore
Shelter Sizing up the phase
Animals must graze

Seeking yonder shower
of influxes

Chizzled
grizzly
Zeplin
Into timeless space
Stars dReam
No
haste

Pace
Mace
Pace

Rattled and dazzled
Shooting in plunder
Casted a spell
ov witch I Am undeR
Translated by fox
EveR so Lost
Mayham at my dooR
Spit on the flooR

Soon death will come
An end to a chapter
Reflox
Damaged
Pantishund
Divided
Fluently
Limited
dispare
Tight
Tension
turlvlant
wildfire
I want death before me
Fighting in haste
Death
Reverse Osmosis
Soon there will be death

THERE were branches
I like bushes

Not Has there been such since
times of moses
Claims to have given soul of
Moses to black man
Heaven forbid

Pandoras box
Don't worry
Soon I will die

I will then not have such witchery
The pains of Moses deep inside of me
A people who cannot travel amongst
themselves freely. Traders and joiners
Let these people trade and join together
for to trade and join will never be
murdered, raped, or bred out of their
spirits
They will often move up in thousands
for reasons rulers of the lands can
make no sense of
For there is a moman prince
Born outside of wholey cult

Dawn mesmorized in stupidity
Flashing outside warning

No female without loosened genitals

Not damaged by a male
Raised as a princess
Not a whore in my eyes
Sell herself cheap
For dramatic cosequence
Sickened squirming in my heart
No true princess
Or even princess caliber
Just lust greeding
Heathens of disgust
Nausiating
To think of them as I sleep

For I am a divided man
Without plan
~~trying~~ to join SYSTEMS
For true unity
Amongst peoples
Unity I feel
Will mean legal maryjuana
For me
My lust greed need
Without is the poor house
I feel Vacant without
In my wizardry sesations
Full of prohibitions

I write when I Am stoned
I think about arguing freedom of speech
for my cannibus intoxicant
For without
Penmenship is much more difficult
I like the fluid motion
that Intoxication pushes along
It is another spirt in my body
It seems Kind and gentle
Often
Not always
There is discipline
It is necessary
Without discipline
Is to escape Love
Discipline is necessary to Survive
Don't lay down and give up for death

It is possible to defeat a common foe
Friend OR Foe

There is no friend
Everyone wants something
I've always wondered what it would be like
A friend becomes family
Family & business
Strange sensation
How to Escape the Irony
Dead leading
Tinder Box
Betrayed
In spirit
To ponder
And wandeR
Lost looking for meaning

No one to stand up
And call the vultures
As such

Dire strength diminishes
with thoughts of desent
Ambushed in satire

So a fool remains a fool
Those who choose to rein over this Kingdom
in tyranny are fools

Who will call their Bluff
or will they Kill
To show selfish nature opennly for all
the world to see

What calls out
CANNIBUS

World starvation

War on plants

The good stuff
Marijuana and opium Are getting tougher to get
More speed and Meth
Poison
often more available
Availability influences market

I AM A CRAZY Man
Taking to a claim
Flying High And Dim
Laking to shine
From A distant place
Wondering of common ground
Leo Lion Clown

Hell on my healz
Spiting wild fire

Death is Heavy
Ever Near
All can hear
Hells Bells
It sells

Pleasant
Dream
To be
Seen

sit
git
Fit
Hit
Down donkey
I hate your
Way
Leave me alone
I just need to
play

Fuck not snot
Begott
Mouth + foot Rot
Who fought
Who tought
Not

So deep inside
Heart of stone

Screams of delight
For none can hear

Terror twisted fantasy
It is so easy
Sleeper locked hold
Down it goes

The body
For my enjoyment
To do as I wish
No one to fight
Just me and the body
Her body

But she is no home
Has Now Roamed
To heaven perhaps
I don't know
I will do As I wish
I don't care
I Am Sick

The judge freed me from the looney bin when I promised to do no harm. I drove straight down to mexico crazy to fit in. I moved into a whore house where the rent was cheaper than the hotel. My two week supply of pills ran out so I went to the pharmacy. I need pills for werewolf syndrome I told him as I showed him my empty bottle. He said he had none and angels began to sing, with demons swearing in the background I began to make ~~movies~~ telepathic movies in my head. TV shows only the spirit knows which of them are mine. Then I saw a street drag race and I crossed the finish line. The cops arrested me asking money to let me go. I said I had none and I was prepared to do time. Realizing I was crazy they let me go telling me to leave town. I got in my truck and off I went heading for the boarder. Toll booths I crossed not stopping to pay a dime. Gas stations filled my tank for free when I gave the Rastaman sign.

www.ingramcontent.com/pod-product-compliance
Lightning Source LLC
LaVergne TN
LVHW080000210926
848481LV00012B/3631
9780615172750